Changes in Matter

Harcourt
SCHOOL PUBLISHERS

Orlando   Austin   New York   San Diego   Toronto   London

Visit *The Learning Site!*
www.harcourtschool.com

Mixing Matter

You can make mixtures with solids.
The solids do not become other things.

You can mix some solids and liquids.
Sugar mixes with water and lemon juice.
You can taste the sugar in the lemonade.

Changing Matter

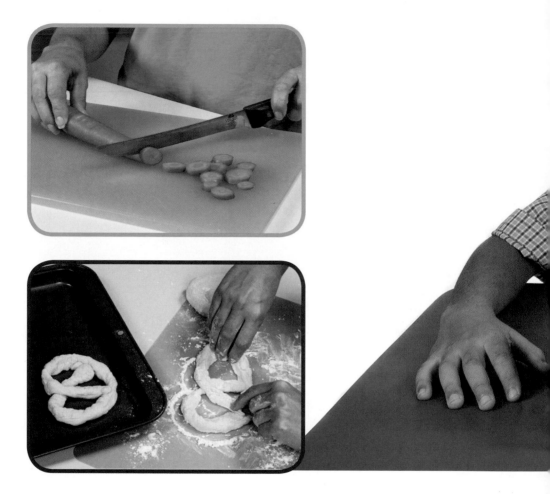

You can change the size of matter.
You can also change the shape of matter.
The mass of the matter does not change.

Cheese is matter.
You can cut cheese into pieces.
The mass of the cheese stays the same.

Freezing

Liquids may freeze and become solids.
This happens when heat is taken away.
All solids have a size and shape.

Melting

Solids, like this ice pop, may become liquid.
This happens when heat is added.
When enough heat is added, the ice melts.
The liquid does not have a shape.

Evaporating and Boiling

Water can change from a liquid to a gas.
This happens when heat makes water boil.
When water boils, evaporation happens.
Water becomes water vapor, a gas.

Condensing

Water vapor can change into liquid water.
This happens when heat is taken away.
Cold water takes heat from the air.
Condensation takes place on the glass.

Burning and Cooking

Burning changes matter into new matter.
It changes wood into ashes and smoke.
They can not change back into wood.

Cooking changes matter into new matter.
Heat can turn a marshmallow brown.
It can change meats and vegetables.
These foods can not change back again.

Vocabulary

burning, p. 10

condensation, p. 9

evaporation, p. 8

mixture, p. 2

water vapor, p. 8